ANNA W. PERRIN

Homesteading For Complete Beginners

A Backyard Guide to Growing Your Food, A Budget-Friendly Path To Keeping Chickens, Canning, Crafting, Herbal Medicine, Back to Basic, and More

First published by Anna w. Perrin 2024

Copyright © 2024 by Anna w. Perrin

All rights reserved. No part of this publication may be reproduced, stored or transmitted in any form or by any means, electronic, mechanical, photocopying, recording, scanning, or otherwise without written permission from the publisher. It is illegal to copy this book, post it to a website, or distribute it by any other means without permission.

Anna w. Perrin asserts the moral right to be identified as the author of this work.

First edition

*This book was professionally typeset on Reedsy.
Find out more at reedsy.com*

Contents

Introduction	1
The world of homesteading	1
Brief overview of what homesteading entails	3
Encouragement For beginners To Embrace	5
Chapter 1: Getting Started	8
Understanding The Homesteading Disposition	8
Evaluate the available space (backyard, balcony, or rooftop)	10
Setting attainable goals for beginners	11
Chapter 2: Growing Your Food	13
The fundamentals of gardening: soil preparation, planting, and care	14
Soil preparation	14
Planting	15
Maintenance	15
Appropriate crops for novices (vegetables, herbs, and fruits).	17
Vegetables	17
Herbs	18
Fruits	18
General Tips for Beginners	19
Seasonal Planting and Crop Rotation.	20
Seasonal Planting	20
Crop Rotation	21
Implementing crop rotation	21
Chapter 3: Canning and Food Preservation.	23
An Introduction To Canning Processes	23
The Two Major Canning Techniques	24

Canning Supplies You Will Need	24
Steps for Successful Canning	25
A step-by-step tutorial for preserving fruits, vegetables, and preserves	27
Preserving Fruit	27
Preservation of Vegetables	28
Making and preserving jams	28
General Tips for Preserving	29
Safety Precautions And Best Practices	30
Safety precautions	30
Best Practices	31
Handling Low Acid Foods	31
Dealing with High Acid Foods	31
If Something Went Wrong	32
Developing a Safe Canning Culture	32
Chapter 4: Keeping Chickens	33
Advantages of backyard hens	33
Choose Chicken Breeds For Beginners	35
Top Beginner-Friendly Breeds	37
Coop Design	37
Feeding	38
Egg Collection	39
Common Health Issues and How to Prevent them	39
Chapter 5: Generate Your Energy	42
Looking At Renewable Energy Options (solar panels, wind turbines)	43
Simple Energy-Efficient DIY Projects	44
Decreased Dependency On The Grid	46
Chapter 6: Crafting and DIY Projects.	48
Creating Crafts From Natural Materials (woodworking, weaving, and pottery)	49
Woodworking with natural materials	49
Woven using natural fibers	50

Pottery With Clay	50
Developing Practical Items For Your Household	51
Decorate Your Home With Handmade Things	53
Chapter 7: Herbal Medicine	56
Growing Therapeutic Herbs	56
Herbal Remedies For Common Illnesses	58
Harvesting and preparing herbal teas, tinctures, and salves	61
Chapter 8: Raising Small Livestock	64
Introduce Honeybees, Bunnies, And Goat	64
Benefits of tiny livestock	66
Basic care and management	67
Healthcare and Veterinary Assistance	68
Handling and Behavior	68
Reproduction and Breeding	68
Pasture and Land Management.	69
Bio security	69
Education & Resources	69
The Best Rotational Grazing	70
Conclusion	72
Share your success	72
Create Something from Your Homestead	73
Encourage newcomers to learn, adapt, and enjoy the experience	74

Introduction

The world of homesteading

I can still remember the day I decided to leave the city. Those giant skyscrapers seemed taller than my dreams and cast long shadows. I wanted sunlight, dirt, and a genuine life. That is when I opted for homesteading which became the beginning of my story as it changed me forever.

I started small, with a patch of land that was more weeds than soil. However as I turned towards the earth and planted seeds with full hope, then I felt a connection to the land that had never been there before to me. It was like whispering abundance whenever the first green sprouts came up.

As time went by, so did I. It taught me about the sun's rhythm and rain's dance. In the summer heat, my tomatoes became red while under harvest moon pumpkins swelled. Canning made me learn how to set the colors of seasons in jars: golden peaches; red strawberries; green pickles Candies et al.

Coming after that were the hens with their clucks, a happy sound that went well with my days. They were giving me eggs even if it was a simple thing but also I used to feel so proud of myself for them. And I made a house where they could stay and lay eggs at any time; this house was not straight, or fancy but it was very strong and full of love.

Energy became my next exploration. I looked up at the solar panels, unable to believe how they got power from the sun enabling that light to get into my house. I became an earth-keeper, ensuring there was less water wastage, composting leftovers as well as watching the size of footprints go down in every new project which was completed.

I then tried crafting which gave me something beautiful out of nature. From willow branches; I wove baskets, from fallen tree limbs; spoons were carved and neighbor's sheep wool knitted into scarves. Every piece showed who I am because these crafts depicted experiences acquired by me.

Perhaps, the spellbinding most of all was herbal medicine. I cultivated plants such as Echinacea and Chamomile and unraveled their secrets. I made teas that would cool sore throats, and balm for cuts or bruises. The earth offered, and I became her lowly scholar.

Over time, my tiny homestead grew into a haven not only for me but also for friends and family who needed rest from the frantic pace of life. In my

Introduction

garden, we ate together, laughed together, and told tales together linking one another through friendships that bound like the roots of plants grown there.

This is how homesteading has been for me; it's about growth, learning, and connection. Every dawn signifies the beginning of an additional chapter while each harvest ends on a happy note; this is an ongoing story. Sitting on my verandah every evening thus seeing stars flashing above me gives me the certainty that this is my real existence.

Brief overview of what homesteading entails

For me, homesteading is a trek back to the roots of self-sufficiency, a way that diverts from the consumer-driven world and goes on to a life of simplicity, sustainability, and satisfaction. It is about embracing the role of the producer not just the consumer; it's about nurturing land and in turn being nurtured.

I delved into this adventure with just an idea and a patch of earth. But with each seed put into the ground, every chick hatched, and all those sealed jars sewed together the fabric of my homestead forming a tapestry woven in the colors of mother earth and nature's rhythm.

Gardening became something I did every day as a form of meditation. I started to read leaf language, listen to wind whispers, and respect the sun's power. Patience and perseverance were taught by the garden as I waited for sprouting leaves to break through the first green shoots indicating future harvests.

Preserving and canning allowed me to capture the essence of each season in jars lined up like precious stones on my pantry shelves. The sweetness of summer, the spice of autumn, the warmth of winter, and the freshness of spring—all were there for the taking, even when the garden lay dormant under a blanket of snow.

Raising chickens was a lesson in the cycle of life. From the gentle chirping of chicks to the proud clucking of hens laying their first eggs, these feathered companions provided more than just food; they offered companionship and a daily reminder of life's simple joys.

Harnessing energy from the sun and wind was an act of defiance against the confines of the grid. It was a declaration of autonomy, a statement that I could power my own life, both literally and metaphorically, with the forces that have fueled the earth since ancient times.

Crafting became a means of transforming the natural resources of my homestead into objects of beauty and practicality. Whether it was repurposing driftwood into a rustic frame or knitting a cozy hat from homemade yarn, crafting was a tangible display of creativity and ingenuity.

Herbal medicine was a rediscovery of age-old wisdom, a way to heal myself and my loved ones with the plants that thrived at my doorstep. Each herb, from soothing lavender to invigorating mint, was a gift from the earth, a natural remedy waiting to be utilized.

As my homestead expanded, so did my sense of community. Neighbors would stop by to exchange seeds, share advice, or simply admire the evolution. We swapped goods and services, yes, but we also exchanged anecdotes, laughter, and encouragement.

Homesteading is not merely a series of chores; it is a way of life, a philosophy, and a decision to live intentionally and purposefully. It is about finding happiness in the toil of your labor, the dirt beneath your fingernails, and the bounty in your harvest. It is about going to bed weary but content, knowing that you have lived a day true to your beliefs and aspirations.

Ultimately, homesteading is about connection—connection to the land, to the food you consume, to the energy you harness, and to the community you establish. It is about actively engaging in your own life, taking charge of your needs, and discovering abundance in what may appear scarce to others. This is the realm of homesteading as I perceive it—a realm where each day presents an opportunity to learn, develop, and flourish. It is a realm I have chosen and one that has embraced me, a realm where the fruits of labor are the most satisfying of all.

Introduction

Encouragement For beginners To Embrace

Setting off on a journey towards self-reliance is a life-changing adventure that can enhance your life in numerous ways. It's a route that leads to increased autonomy, a stronger bond with the natural world, and a profound sense of achievement. If you're a novice contemplating this way of life, allow me to enlighten you on the countless advantages and pleasures that lie ahead.

Self-reliance is not merely about cultivating your produce or generating your power; it's a mindset, a belief system that enables you to seize command of your life. It's about making deliberate choices that resonate with your principles and ambitions.

By opting for this course, you're opting to live purposefully, to diminish your ecological impact, and to make a positive contribution to the world around you. Picture waking up each day to the melody of birds, rather than the harsh sound of an alarm clock. You step outdoors, and there's a garden awaiting you, brimming with vegetables and herbs that you've nurtured yourself.

There's a tranquility that comes from knowing you're fostering life, that you're part of a cycle that nourishes not only your body but your spirit. As a beginner, the idea of starting a homestead may seem overwhelming, but keep in mind, every expert was once a novice. Begin small, with a few pots of herbs on a windowsill or a solitary raised bed in your backyard.

Rejoice in every new shoot, every bloom, and every harvest, regardless of how modest. These minor triumphs will bolster your confidence and abilities, and before long, you'll discover yourself expanding your garden, experimenting with new methods, and delving into new facets of homesteading.

- Canning and storing your crop is a great method to stretch the bounty of your garden all year. There's something magical about opening a jar of homemade tomato sauce in the middle of winter, giving you a taste of summer when everything else is cold and gray. It serves as a reminder of nature's cycles, your hard work, and the resilience that comes with self-sufficiency.

- Raising hens or keeping bees may seem like a big step, but these animals can provide a lot of joy and life to your homestead. They supply more than just eggs and honey; they connect us to the web of life and serve as a daily reminder of all living beings' interdependence.
- Creating your electricity with solar panels or wind turbines is a powerful step toward independence. It expresses that you are more than just a resource consumer; you are also a steward of the world. Each kilowatt-hour you generate is a step away from fossil fuels and a contribution to a cleaner, greener planet.
- Crafting and DIY projects enable you to show your creativity while meeting your necessities. Whether it's knitting a warm sweater, building a bookshelf, or creating soap, these activities honor the human spirit and the ability to create beauty and utility from simple materials.
- Herbal medicine offers you a world of natural healing, allowing you to care for yourself and your loved ones using the plants that grow around you. It's old wisdom that remains relevant today, a method to draw into nature's healing power while reducing your need for medications.

```
As you learn more about homesteading, you'll find a community of
like-minded folks willing to share their knowledge, seeds, and
stories. Homesteading is not a solo undertaking; it is a global
network of relationships and assistance.
The difficulties you will encounter along the way are part of the
trip. Each difficulty surmounted is a lesson learned,
demonstrating your resilience and inventiveness. There will be
mistakes and victories, and each experience will get you closer to
understanding what it means to be self-sufficient.
Adopting a self-sufficient lifestyle alters not just your way of
life, but also your perspective on the world. You'll observe small
seasonal variations, how the light fluctuates throughout the year,
and the patterns of the birds and insects. You'll have a sense of
belonging to the land, as well as a connection to a specific
location and way of life.
```

Finally, self-sufficiency is about freedom—the ability to make decisions that are healthy for you, your community, and the environment. It is about living a life that is genuine, meaningful, and deeply fulfilling. So, take the first step, plant the first seed, and go on a journey that will change your life in ways you cannot yet conceive. The world of homesteading welcomes you with open arms and limitless possibilities.

Remember, this is a thousand-mile journey that starts with one step. Take that step now and let the experience begin.

Chapter 1: Getting Started

Understanding The Homesteading Disposition

Chapter 1: Getting Started

The homesteading philosophy emphasizes a self-sufficient and sustainable existence. It is a style of life in which you are responsible for your requirements, from cultivating your food to producing electricity for your home. Here's a straightforward explanation of this mindset:

Homesteaders have a strong connection to nature. They work with the soil rather than against it, and they enjoy the natural cycles of food production, harvesting, and preservation.

Self-reliance is the foundation of homesteading. It involves relying on your abilities and efforts to meet your demands. Homesteaders take satisfaction in doing things on their own, whether it's mending a leaky faucet or baking homemade bread.

Homesteaders try to develop a sustainable lifestyle over time. This includes conserving resources, reducing waste, and, in many cases, living off-grid or with a low environmental impact.

Community and Sharing: Self-reliance is important, but so is community. Homesteaders frequently exchange expertise, seeds, and harvests with their neighbors. They form support networks and collaborate to learn.

Homesteading needs a variety of talents, including gardening, construction, and animal husbandry. Homesteaders constantly learn and adapt, whether through books, courses, or trial and error.

Homesteading prioritizes simplicity. It's about removing the unneeded and focusing on what's genuinely important—health, family, community, and the environment.

Resilience: Homesteaders are problem solvers. They are resilient and resourceful, seeking answers to problems and recovering from failures.

The homesteading philosophy is based on the concepts of self-reliance, sustainability, and community. It's a wonderful way of life, with every day presenting fresh opportunities to grow, create, and prosper.

Evaluate the available space (backyard, balcony, or rooftop)

Assessing your available area, whether in the backyard, balcony, or rooftop, is an important first step in the homesteading process. It's about realizing your space's full potential and transforming it into a productive and sustainable environment. Here's how to handle this assessment:

1. Measure the measurements of your space. This will offer you a clear picture of the space you have to work with. Measure the length and width of your backyard to calculate square footage. Consider the vertical area for potential hanging plants or vertical gardens when designing balconies and rooftops.
2. Evaluate Sunlight Exposure: Assess the quantity of sunlight your space receives during the day. Most food plants need at least 6 hours of direct sunlight to flourish. Take note of any tall buildings, trees, or other structures that may create shadows and block sunlight.
3. Assess Soil Quality (Backyards): Assess the soil quality in your backyard. Is it sand, clay, or loamy? You can use a simple soil test to detect the pH and nutrient levels. Good soil is the cornerstone of a successful garden, so consider adding compost or other organic matter as needed.
4. Evaluate Water Access and Drainage: Ensure a convenient water supply is nearby. To avoid water logging, examine the drainage system on balconies and rooftops. Plants require appropriate water to develop, but too much can be just as damaging as too little.
5. Understand the Climate and Micro climate: Your local climate affects what you can grow and when. Your individual space may also have a micro climate, which includes temperature, wind, and humidity fluctuations. Understanding these variables will assist you in selecting the appropriate plants and protecting them from adverse weather conditions.
6. Plan for Accessibility: Ensure your location is easily accessible for regular upkeep. Make sure you can access all regions comfortably when planting, watering, and harvesting. If mobility is a concern, consider

raised beds or container gardening.
7. Consider Aesthetics: Your space's visual appeal is equally vital as its utility. Plan your homestead to be both productive and aesthetically beautiful. Create an eye-catching layout by combining plant heights, colors, and textures.
8. Be aware of potential wildlife and pests in your space. While some critters are useful, others might harm your plants. Consider using protective measures such as fencing or netting if necessary.
9. Set realistic goals within your environment. Do not overcommit or overcrowd your region. Start small, gain experience, and eventually develop your homestead.
10. Develop a Plan: After assessing your space, write a detailed plan. Draw a plan for where each plant or feature will go, taking into account its mature size and sunlight and water requirements.

By thoroughly examining your place, you may optimize its potential and develop a successful homestead, regardless of size. Remember that homesteading is about making the most of your resources while also enjoying the process of growing and learning.

Setting attainable goals for beginners

Setting realistic goals is a critical component of successful homesteading, particularly for beginners. It is about striking a balance between ambition and practicality, dreams and realities. Here's how you may create realistic goals as you begin your homesteading journey:

Start with little, doable initiatives that don't demand substantial time or resources. This could include growing a small vegetable garden, starting a compost pile, or learning how to bake bread. Small victories will help you gain confidence and skills.

Educate Yourself: Learn about homesteading principles. Read books, attend courses, and meet with seasoned homesteaders. Knowledge is power, and the more you know, the more effectively you can plan and carry out your objectives.

Evaluate your available resources, including space, time, and money. If you have a small backyard, don't expect a large herd of livestock. If you have limited time, pick low-maintenance projects.

Be patient.

Understand that homesteading skills require time to learn. You will not become a skilled gardener or carpenter overnight. Allow yourself the freedom to learn and develop at your own pace.

Prioritize long-term sustainability while setting goals. This involves not overburdening yourself or your resources. Plan for tasks that you can sustain over time, providing a continuous path to self-sufficiency.

Prioritize:

Decide what is most essential to you and your homesteading goals. Is it about achieving food independence, increasing energy efficiency, or decreasing waste? Prioritize goals that are consistent with your basic values.

Break down goals into smaller, achievable milestones. Instead of "start a garden," set a goal like "prepare garden beds by March, plant seeds in April, and harvest lettuce in June." This makes your goals more measurable and attainable.

Be flexible and change your goals as you learn more about homesteading and yourself. What works for one individual may not work for another, and that's fine. Flexibility is essential in determining what works best for your specific situation.

Celebrate your achievements, no matter how minor. Every stride forward represents progress, and acknowledging your efforts will keep you motivated.

Regularly reflect on your goals and progress. What works? What is not? Use these reflections to adjust your goals and strategies.

Setting realistic goals lays the groundwork for a rewarding and sustainable homesteading lifestyle. Remember: homesteading is a marathon, not a sprint. Enjoy the trip, learn from the trials, and enjoy each step toward self-sufficiency.

Chapter 2: Growing Your Food

The fundamentals of gardening: soil preparation, planting, and care

Gardening is a joyful pastime that helps us connect with nature and gives us a sense of success. Understanding the fundamentals of gardening is critical for growing flowers, veggies, or herbs. Here's a comprehensive guide to soil preparation, planting, and care to help you grow a thriving garden.

Soil preparation

Soil is the foundation of all gardens. It is where your plants will take root and receive nutrients and water. There are numerous steps to prepare your soil:

1. **Testing the soil:**

Before you begin, you need to know what type of soil you have. A soil testing kit is available from your local garden center or extension office. This will tell you about the pH and nutritional content of your soil, which is critical for plant growth.

2. **Clearing the area:**

Remove any weeds, rocks, or trash from the place you intend to plant. Weeds can compete with your plants for nutrients and water, so begin with a clean slate.

3. **Tilling the soil:**

Use a spade or tiller to turn the soil. This aerates the soil and promotes root growth. It also aids in the incorporation of any necessary adjustments.

4. **Adding amendments:**

Based on your soil test results, you may need to apply amendments to improve the soil. If your soil is excessively acidic, you can add lime; if it is too alkaline, sulfur may be required. Compost is usually always a beneficial addition since it increases nutrients and improves soil structure.

5. **Leveling and Raking:**

After you've added the amendments, level the soil with a rake to ensure an even surface. This also helps to remove any large clumps of soil or debris.

Planting

After you've prepped your soil, it's time to plant.
   1. **Choosing Plants:**
   Choose plants that are appropriate for your environment and the amount of sunlight that your garden gets. Local garden centers can provide information on the best plants for your location.
   2. **Spacing:**
   Pay close attention to the recommended spacing for each plant. Overcrowding can result in inadequate air circulation and competition for resources, making plants struggle.
   3. **Digging holes:**
   For each plant, dig a hole just large enough to hold the root ball. The top of the root ball should be level with the soil surface.
   4. **Planting:**
   Gently remove the plant from its container, loosening any tightly twisted roots. Place the plant in the hole and fill in with dirt, gently pressing down to remove air pockets.
   5. **Watering:**
   Water your new plants thoroughly to settle the soil and deliver moisture to the roots.

Maintenance

Regular upkeep is essential for a healthy garden.
   1. **Watering:**
   Water your plants as needed, focusing on the plant's base rather than the foliage to prevent illness. The amount and frequency with which you water your plants will be determined by your climate and their requirements.
   2. **Mulching:**

Apply mulch around your plants to help retain moisture, regulate soil temperature, and inhibit weed growth.

3. **Weeding:**

Keep an eye out for weeds and get rid of them quickly. Weeds may quickly take over a garden if not controlled.

4. **Pruning:**

Pruning your plants removes dead or unhealthy foliage and promotes healthy development. Different plants require different trimming techniques, so research the best procedures for each type.

5. **Fertilizing:**

Feed your plants with the appropriate fertilizer to ensure they receive the nutrients they require. Over-fertilization can be detrimental, so follow the directions on the fertilizer package.

6. **Pest and Disease Control:**

Keep an eye out for symptoms of pests or illness in your plants. Early discovery is critical to addressing these concerns. Use organic or chemical controls as needed, adhering to all safety rules.

7. **Seasonal Care:**

Be conscious of your garden's demands throughout the seasons. Some plants may require frost protection, while others may necessitate fall pruning.

By following these simple procedures for soil preparation, planting, and care, you'll be well on your way to developing a beautiful and productive garden. Remember that gardening is a learning process, with each season presenting new problems and chances for improvement. Enjoy your adventure!

Appropriate crops for novices (vegetables, herbs, and fruits).

Choosing the correct crops for your first garden is like laying the groundwork for a structure; it's critical to success. As a beginner, you should choose plants that are tolerant, take little maintenance, and produce a plentiful yield. Here's a complete guide to selecting the best veggies, herbs, and fruits for your first gardening experience.

Vegetables

Vegetables are the foundation of any garden, providing fresh vegetables for your table. Here are some top choices for beginners:

> 1. Lettuce:
> - Why: Lettuce grows swiftly and does not require much area. It's also adaptable, with numerous options to pick from.
> - How to Plant: Grow in loose soil with partial to full sunlight. Keep the soil wet.
> - Harvest time: Leaf lettuce is ready in 40-45 days, while head lettuce takes 80-95 days.
> 2. Radishes:
> - Why: Radishes are among the fastest-growing vegetables, ready for harvest in as little as 20-30 days.
> - How: Sow directly into loose, sandy soil that receives full sun exposure.
> - Harvest: The quick turnaround makes them appealing to new gardeners.
> 3. Green Beans:
> - Why: They are resilient and require little upkeep. Bush types are particularly suitable for beginners.
> - How: Needs full light and well-drained soil. Direct sow following the recent frost.
> - Harvest time ranges between 50 and 65 days, depending on the

cultivar.

Herbs

Herbs are ideal for novices because they are simple to grow and may be utilized in a wide range of meals.

   1. *Basil:*
   - Why: Basil is a warm-weather herb that works well in pesto and as a fresh accent to salads.
   - How: Plant in a sunny location once the threat of frost has passed. Basil enjoys warmth and regular watering.
   - Harvest the leaves as needed to encourage the plant to become bushier.
   2. *Mint:*
   - Why: Mint is extremely hardy and can grow invasive if not controlled.
   - How to: Grow in a pot to keep it from taking over your garden. It prefers wet soil with some shade.
   - Harvest: Cut the sprigs as needed. The more you harvest, the larger it grows.
   3. *Parsley:*
   - Why: Parsley is a biennial herb full of flavor and minerals.
   - How to Plant: Grow in rich, moist soil in full sun or partial shade.
   - Harvest: Remove the outer leaves first, leaving the center leaves to continue developing.

Fruits

Fruit plants can be more difficult, but there are still solutions available for beginners.

   1. *Strawberries:*
   - Why: Strawberries are a delightful pleasure that can be cultivated in either the ground or in containers.

- How: Plant in well-drained soil that receives lots of sunlight. Make sure they have room to spread.
- Harvest: Typically ready in late spring to early summer.

2. Raspberries:
- Why: Raspberries are perennials that produce fruit for years.
- How to Plant: Choose a sunny spot with well-drained soil. They need help as they grow.
- Harvest: Fruit is ready when it can easily be separated from the plant.

3. Tomatoes:
- Why: Tomatoes, despite being technically classified as a fruit, are a vegetable garden mainstay.
- How: Support requires staking or caging. Plant in a sunny area with healthy soil.
- Harvest when they are firm and completely colored.

General Tips for Beginners

- Begin Small: Do not overwhelm yourself. Begin with a few plant varieties and progress as you develop confidence.
- Understand Your Zone: Determine your climate and select plants that will thrive in your location.
- Soil Matters: Invest in quality soil. It is the basis for your garden's health.
- Water Wisely: Learn about your plant's watering needs. Over watering can be as dangerous as under watering.
- Label everything: Keep track of what you're planting and where. It will make maintenance simpler.
- Be Patient: Plants need time to grow. Don't get discouraged if things don't happen immediately.
- Enjoy the Process: Gardening is a learning opportunity. Enjoy the journey as much as the destination.

To summarize, the key to selecting the correct crops as a novice is to choose ones that are simple to grow and manage, bring immediate gratification, and are adaptive to your gardening conditions. Lettuce, radishes, and green beans are wonderful vegetable choices; hardy herbs include basil, mint, and parsley; and fruit plants like strawberries, raspberries, and tomatoes provide a pleasant reward for your efforts. With these crops, you will have a successful and joyful first gardening season. Happy planting.

Seasonal Planting and Crop Rotation.

Seasonal planting and crop rotation are important gardening and agricultural methods that improve soil health and productivity while also controlling pests and illnesses. Here's a detailed explanation of both concepts.

Seasonal Planting

Seasonal planting is the practice of seeding crops at the most appropriate times of the year for growth. This takes into account plants' natural life cycle and the environmental conditions necessary for maximum development.
- Understanding Climate: Different crops grow in different seasons; cool-season crops like lettuce and spinach prefer the milder temperatures of spring and fall, whereas warm-season crops like tomatoes and peppers prefer the heat of summer.
- Maximizing Harvests: By coordinating planting schedules with the seasons, gardeners may ensure a consistent crop all year. For example, planting cool-season crops in early spring and again in late summer can result in two harvests from the same area.
- Resource Efficiency: Seasonal planting conserves water and other resources by providing plants with natural rainfall and sunlight when they are most needed.

Crop Rotation

Crop rotation is the process of cultivating various crops in the same place over several seasons or years. This technique prevents nutrient depletion, disrupts pest and disease life cycles, and preserves soil structure and fertility.

- Breaking Pest Cycles: Numerous pests and illnesses are crop-specific. Rotating crops eliminates their preferred host, lowering their numbers and damage.

- Nutrient Management: Each crop has different nutrient requirements. Legumes, for example, fix nitrogen in the soil, whereas leafy greens may be high nitrogen consumers. Rotating these crops helps to balance soil fertility.

- Soil Structure: Each plant has unique root systems and growth habits. Rotating deep-rooted and shallow-rooted crops can help to maintain soil structure and minimize compaction.

Implementing crop rotation

To successfully execute crop rotation, consider the following:

> - Family Groupings: Rotate crops according to their botanical families. For example, do not grow tomatoes after peppers because they are in the same family and are prone to comparable diseases.
>
> - Rotation Plan: Develop a three to four-year rotation plan to guarantee that crops from the same family are not planted in the same location too often.
>
> - Keeping detailed records of what is planted where and when allows you to track the rotation schedule and make informed decisions about future plantings.

Advantages of combining both practices:
Combining seasonal planting with crop rotation can result in a more productive and sustainable garden. It makes the best use of available space

throughout the year while also keeping the soil healthy and productive. This technique has the potential to increase yields, require less fertilizer and pesticide inputs, and create a more resilient garden environment.

In summary, seasonal planting and crop rotation include collaborating with nature to produce a harmonious and sustainable growing environment. Gardeners who understand and execute these techniques can enjoy abundant harvests, healthier plants, and a more vibrant and sustainable garden.

Chapter 3: Canning and Food Preservation.

An Introduction To Canning Processes

Canning is a time-honored technique for preserving seasonal harvest. Whether you've harvested a surplus from your garden or found a great deal at the farmers' market, canning allows you to savor the flavors of fresh produce all year. Here's a detailed canning guide broken down into easy-to-follow steps.

Understanding canning:

Canning is the method of preserving food in airtight containers, usually glass jars, that may be kept at room temperature. The goal is to establish an environment in which bacteria, yeasts, and molds cannot grow, ensuring that the food is safe and tasty for months or years.

The Two Major Canning Techniques

1. Water Bath Canning:

 - Ideal for high-acid foods such as fruits, tomatoes, pickles, jams, jellies, and fruit juice.

 - Process: Food-filled jars are placed in a water-filled canner, making sure they are thoroughly submerged. The water is then brought to a boil, and the jars are processed for the amount of time recommended by the recipe.

2. Pressure Canning:

 - Ideal for Low-acid foods such as vegetables, meats, and poultry.

 - Procedure: Jars are placed in a pressure canner with a set amount of water (often 3 quarts). The lid is secured, and the canner is heated to generate pressure. Once the desired pressure has been achieved, the jars are processed for the period stated in the recipe.

Canning Supplies You Will Need

- Canners include a water bath caner for high-acid foods and a pressure canner for low-acid foods.

 - Jars: Mason jars are the standard for canning and come in a variety of sizes to meet your needs.

 - Lids and Rings: While jars can be reused, lids are normally used once to

ensure a proper seal.
  - Jar Lifter: A tool for carefully removing hot jars from the canner.
  - Canning Funnel: Allows you to fill jars without spilling.
  - Bubble Remover: A plastic or wooden instrument for releasing air bubbles trapped in a jar.
  Canning salt is a non-iodized salt used to preserve food

Steps for Successful Canning

1. Prepare your jars:
  - Clean jars, lids, and rings with hot, soapy water.
  - Boil the jars for 10 minutes or use the sanitize setting on your dishwasher.
  - Keep jars warm until they are ready to be filled to avoid them breaking when filled with hot food.
2. Prepare your food.
  - When preparing your food for canning, use a tried-and-tested recipe. This could include peeling, chopping, frying, or seasoning.
  - Follow the recipe's directions for whether to hot pack (load jars with hot food) or cold pack (fill jars with raw food).
3. Fill your jars:
  - Fill jars with a canning funnel, leaving the recipe's necessary headspace (the area between the food and the top of the jar).
  - Run a bubble remover or a non-metallic spatula around the jar's inside edge to eliminate any air bubbles.
  - Wipe the jar's rim with a clean, wet cloth to ensure a proper seal.
4. Seal your jars:
  - Place the lid on the jar, making sure the sealing compound is in contact with the rim.
  - Screw the band on until it is finger-tight; do not overtighten.
5. Process your jars:
  - Use a jar lifter to place the jars into the canner.
  - For water bath canning, make sure the jars are covered by at least 1 inch of water. Bring to a boil, then process for the time stated in the recipe.

- For pressure canning, follow the manufacturer's instructions to bring the canner to the proper pressure, then process for the recipe's recommended time.

6. Cool your jars:
- After processing is complete, turn off the heat and let the jars sit in the canner for 5 minutes.
- Remove the jars with the jar lifter and lay them on a towel or cooling rack, leaving space between them.
- Let the jars cool for 12-24 hours without disturbing them.

7. Check your seals:
- Once the jars have cooled, verify the seals by pressing the center of the lid. If it does not pop back, the jar is sealed.
- Remove the bands and try to take the lid off with your fingertips. If it does not fall off, the seal is intact.
- Label your jars with the contents and date of canning.

8. Store your jars:
- Keep sealed jars in a cold, dark place.
- If a jar fails to seal, refrigerate it and use the contents within a few days.

Safety Tips:
- To ensure safety, always use tested recipes from trusted sources.
- Follow the processing times and techniques specified in the recipe exactly.
- If you are unsure whether food has been canned securely, it is best to avoid consuming it.

Canning is both an art and a science. It necessitates attention to detail, cleanliness, and respect for the procedure. But the benefits are great: jars lined up like gems, loaded with seasonal delicacies, and ready to be savored whenever you want.

Chapter 3: Canning and Food Preservation.

A step-by-step tutorial for preserving fruits, vegetables, and preserves

Preserving your garden harvest or market treasure is a delightful way to savor the flavors of fresh produce throughout the year. Here's a step-by-step guide to preserving fruits, vegetables, and preserves, so you can enjoy the flavors of each season no matter what time of year.

Preserving Fruit

1. Selection and Preparedness:
   - Select ripe, undamaged fruits for optimal flavor and preservation quality.
   - Rinse the fruits thoroughly with running water to remove any dirt or residue.
   - Peel, pit, or core the fruits as needed, then cut them into consistent pieces to ensure even storage.
2. Blanching (for certain fruits):
   - Blanching involves quickly boiling fruits to kill enzymes that cause deterioration.
   - Immerse the fruit pieces in boiling water for 30 seconds to a minute before rapidly transferring them to icy water to stop the cooking process.
3. Syrup Packaging:
   - Make syrup by dissolving sugar in water. The syrup's concentration can range from light to heavy, depending on the sweetness of the fruit and your personal preferences.
   - Place the prepared fruit in canning jars and top with hot syrup, leaving a suitable headspace.
4. Canning:
   - For highly acidic fruits, use the water bath canning process.
   - Process the jars in boiling water for the duration provided in a reliable recipe, adjusting for altitude as needed.
5. Cooling and storage:
   - After processing, remove the jars from the water and allow them to cool

for 12-24 hours.
   - Check the seals, mark the jars with the contents and the date, and keep them in a cool, dark area.

Preservation of Vegetables

1. Selection and Preparedness:
   - Harvest veggies at their peak freshness for the best flavor and nutritional value.
   - Thoroughly clean the vegetables, then chop or slice them as desired.
   2. Blanching:
   - Blanching most veggies before preserving them helps to maintain their color, flavor, and texture.
   - Blanch the vegetables in boiling water or steam, then quickly cool them in icy water.
   3. Pickling is optional.
   - Pickling vegetables can add taste and help preserve them.
   - Make a vinegar-based pickling solution with spices and herbs; pour it over the veggies in jars.
   4. Canning:
   - To assure the safety of low-acid veggies, can they be under pressure?
   - Process the jars according to the pressure and time advised by a reliable source.
   5. Cooling and storage:
   - Let the jars cool gradually after processing.
   - Check the seals, label the jars, and keep them in a cool, dark place.

Making and preserving jams

1. Fruit preparation:
   - Choose high-quality, ripe fruit for jam-making.
   - Clean, peel, and smash the fruit, or chop it into little pieces.
   2. Cooking the jam:

- In a big pot, combine the prepared fruit, sugar, and lemon juice (if needed).
- Cook the mixture on medium-high heat, stirring regularly to avoid sticking and burning.

3. Test for Doneness:
- To determine whether the jam is ready, place a small quantity on a cool plate. If it wrinkles when pushed with a finger, it is finished.
- Another option is to use a thermometer; most jams are set at approximately 220°F (104°C).

4. Canning:
- While the jam is still hot, transfer it to sterilized canning jars, leaving the necessary headspace.
- Clean the rims, add the lids and rings, and process in a water bath canner for the time provided in a dependable recipe.

5. Cooling and storage:
- After processing, allow the jars to chill for 12 to 24 hours.
- Check the seals, label the jars, and keep them in a cool, dark location.

General Tips for Preserving

- Use Proper Equipment: Make sure you have all of the necessary canning supplies, such as jars, lids, canners, and utensils.
- utilize Trusted Recipes: Always utilize recipes from reputable sources to ensure food safety and quality.
- Maintain Cleanliness: Keep your workspace and equipment clean to avoid infection.
- Label Everything: Keep track of your inventory by clearly labeling your preserved goods with the contents and date.
- Store Properly: Keep your preserved goods in a cool, dark place to maintain quality and extend shelf life.

By following these methods, you can reliably preserve fruits, vegetables, and jams, resulting in a cupboard full of handcrafted treats that capture the essence of each season. Whether you're an experienced preserver or a beginner, the process is a rewarding way to connect with your food and enjoy the results all year.

Safety Precautions And Best Practices

Safety is the most important consideration when canning. Preserving food in airtight containers for extended periods necessitates great attention to detail and strict adherence to safety requirements to avoid foodborne infections, including botulism. Here's a thorough guide to the safety measures and best practices for canning:

Understanding the Risk

Botulism is a severe illness caused by a toxin generated by the bacterium Clostridium botulinum, which can flourish in contaminated canned foods[1].

- Spoil: Improper canning can cause spoiling, which not only spoils the food but also renders it dangerous to consume.

Safety precautions

1. Always use USDA-approved modern canning processes and recipes.
2. Before canning, check your jars, lids, and equipment for signs of damage or wear.
3. To prevent hazardous microorganisms, sterilize all jars, lids, and canning utensils before use.
4. Keep your workspace clean to prevent cross-contamination throughout the canning process.
5. For high-acid foods, use a water bath canner, while low-acid foods require a pressure canner.
6. Check Seals: After canning, ensure that all jars are sealed properly. If a jar hasn't been sealed, refrigerate it and utilize the contents quickly.
7. To store canned foods properly, keep them cold, dry, and away from direct sunlight.

Chapter 3: Canning and Food Preservation.

Best Practices

1. Choose fresh, high-quality fruits and vegetables that are free of infections and blemishes.
2. To improve the safety of low-acid foods like tomatoes, add lemon juice or citric acid.
3. Leave enough headspace in each jar for expansion during processing.
4. Correctly process jars for the appropriate period, adjusting for altitude as needed.
5. Allowing jars to cool gradually after processing helps minimize breakage and ensures effective sealing.
6. Keep track of your inventory by labeling each jar with the contents and date of canning.
7. Check your pressure canner's dial gauge annually to maintain accurate readings.

Handling Low Acid Foods

- Pressure canning: To eliminate botulism spores, low-acid foods must be treated at high temperatures.
   - No shortcuts: When it comes to low-acid foods, you should never take shortcuts. The risk of botulism is too great.

Dealing with High Acid Foods

Water bath canning is a safe way to prepare high-acid foods.
   - Keep the water at a rolling boil throughout the processing period.
   After canning
   - To test the seal, press the center of the lid to ensure it does not pop back. A proper seal is essential for safety.
   - Proper Storage: Keep canned foods in a cool, dark place to preserve quality and shelf life.

If Something Went Wrong

- Do Not Taste: If you suspect a canned item has rotted or the seal has failed, do not consume it. Dispose of it safely.

- To assure safety, boil home-canned veggies for 10 minutes before consumption.

Continuous Learning

- Stay informed: Canning techniques and safety rules are constantly evolving. To stay informed, examine resources such as the USDA's Complete Guide to Home Canning or your state's County Extension Service.

Developing a Safe Canning Culture

- Educate Others: Share safe canning techniques with friends and family who are also capable.

- Community Resources: Use community resources such as canning workshops and extension agencies for advice and assistance.

By following these safety precautions and best practices, you can ensure that your home-canned foods are both delicious and safe to eat. Remember, canning is a science, and precision is essential.

Chapter 4: Keeping Chickens

Advantages of backyard hens

Backyard chickens have grown in popularity as more individuals discover the several advantages of keeping their flock. Here's a complete look at the benefits of raising chickens in your backyard:

1. Fresh and Nutritious Eggs: The primary advantage of rearing backyard chickens is the availability of fresh eggs. Home-raised hens produce more nutritious eggs with higher levels of vitamin E and omega-3 fatty acids compared to store-bought ones. The yolks will have a richer flavor and a more brilliant hue.
2. Improved Garden Health: Chickens can considerably improve the health of your garden. They naturally plow the soil when foraging, which helps aerate it. Their droppings are a good fertilizer, rich in nitrogen, phosphate, and potassium. This results in healthier plants and more productive gardens.
3. Pest Control: Chickens are natural pest controllers. They consume insects, grubs, and even small rodents, which can help keep your garden and yard free of pests that would otherwise harm your plants or become a nuisance.
4. Sustainability: Raising backyard chickens contributes to a sustainable food chain. You lower your carbon footprint by avoiding the transportation and packaging associated with store-bought eggs. You can feed your chickens kitchen leftovers to reduce food waste.
5. Educational Value: Chickens provide a hands-on learning opportunity for families with children. Children learn responsibility through daily care routines and get an awareness of where their food originates from. It provides practical teaching in biology, agriculture, and nutrition.
6. Mental Health Advantages: Caring for chickens can be therapeutic. The regularity of feeding, cleaning, and collecting eggs instills a sense of purpose and accomplishment. Watching chickens can provide enjoyment and relaxation.
7. Financial Benefits: Setting up a coop and buying feed may require an initial expenditure, but raising hens can result in long-term cost savings. You'll have a consistent supply of eggs, and if you have excess, you can

sell them for extra money.
8. Organic Eggs: Backyard chickens provide control over their diet, making them an ideal option for those who prioritize organic food. Providing an organic diet for your chickens can result in organic eggs at a lower cost than store-bought eggs.
9. Community Building: Participating in poultry chores or sharing eggs helps foster a sense of community. Community relationships can be strengthened by trading gardening ideas for eggs or inviting friends over to hear about homesteading efforts.
10. Animal Welfare: Raising your hens ensures they live in a humane environment. Unlike commercial egg farms, you may provide your chickens with freedom to wander, fresh air, and a healthy life.
11. Self-Sufficiency: Raising chickens leads to a more self-sufficient lifestyle. Having less reliance on grocery stores for basic requirements might be especially reassuring during times of economic or societal uncertainty.

To summarize, backyard hens provide numerous benefits that go far beyond fresh eggs. They contribute to a better garden, promote sustainability, provide educational opportunities, improve mental health, can result in financial savings, provide a supply of organic eggs, develop community, ensure animal care, and increase self-sufficiency. These feathered friends can be a welcome addition to your home, bringing a touch of farm life into your backyard.

Choose Chicken Breeds For Beginners

Choosing the appropriate chicken breed is critical for beginners because it can significantly impact your experience with backyard poultry raising. Here's a complete guide to assist you choose the best chicken breeds if you're just getting started:

1. Consider Your Climate: Each breed has distinct tolerances for heat and cold. For example, the Australorp is recognized for being tough in chilly areas but may not do as well in extreme temperatures.. Conversely,

species like the Minorca are well-suited to warmer regions. Choose a breed that can thrive in your local weather conditions.
2. Determine Your Goals: Do you raise hens for eggs, meat, or as pets? Some breeds, including the Rhode Island Red and Leghorn, are outstanding egg layers, while others, such as Broilers, are better for meat. Consider Sussex or Buff Orpingtons as dual-purpose breeds for eggs and meat.
3. Space Considerations: The amount of available space will impact your choice. Some breeds, such as the Plymouth Rock and Orpington, may thrive in small spaces, while others may need more area to roam.
4. Determine Your Attention Commitment: Some breeds require more maintenance and attention. Buff Orpington hens are self-sufficient and easy to care for, making them ideal for those seeking low-maintenance options.
5. Egg Production: If replacing grocery store eggs is a priority, choose high-laying breeds. The Australorp and Buff Orpington have strong egg-laying capacities, averaging 250 and 200-280 eggs annually, respectively.
6. Temperament: For families with youngsters or those seeking an engaging experience, selecting a friendly breed is essential. Australorp and Buff Orpington breeds are ideal for families with young children because of their gentle temperaments.
7. Broodiness: Some breeds, such as the Buff Orpington, are known to be broody, indicating a strong desire to sit on and hatch eggs. Depending on your want to produce chicks, this can be advantageous or disadvantageous.
8. Choose hardy breeds that can withstand common poultry diseases. Rhode Island Red and Sussex are hardy breeds suitable for beginners.
9. Appearance: Although not the most crucial criterion, the appearance of the hens may be important for you. Australorps have striking black plumage with purple and green tints, making them a beautiful addition to any backyard.
10. Use resources such as the USDA's Complete Guide to Home Canning, local extension offices, and reliable websites to learn about different

breeds and their needs.

Top Beginner-Friendly Breeds

Australorps are friendly, have good egg layers, and are hardy in colder temperatures, whereas Buff Orpingtons are docile, and have good egg layers, and broody, making them appropriate for cold settings.[1] Rhode Island Red is a laid-back, hardy breed that produces wonderful eggs. Sussex is a dual-purpose, friendly, and hardy breed. Plymouth Rock is friendly and adaptable to limited places.

Finally, choosing the correct chicken breed requires taking into account your environment, goals, available space, time commitment, egg production requirements, temperament, broodiness, hardiness, and personal preferences. By carefully examining these aspects and selecting beginner-friendly breeds, you may ensure a rewarding and successful backyard chicken-keeping experience.

Designing and feeding the coop, as well as collecting eggs

Designing a chicken coop, establishing a feeding schedule, and implementing an effective egg collection system are all necessary components of successful backyard chicken rearing. Here's a complete guide to help you understand these concepts clearly and comprehensively.

Coop Design

A well-designed chicken coop is safe, comfortable, and easy to maintain. Here are the essential factors to consider:

1. Size: Make sure that each chicken has at least 2-3 square feet of floor space within the coop and 8-10 square feet in the outdoor run.
2. Ventilation: Proper ventilation is essential for good air quality and temperature control. Include vents or windows near the roof to release hot air and promote cross-ventilation without causing draft.
3. Insulation: Insulate the coop to keep the chickens safe from harsh

temperatures. Use materials that will make the coop warm in the winter and cool in the summer.
4. Predator Protection: Protect the coop from predators with a strong structure, hardware cloth instead of chicken wire, and secure locks.
5. Nesting Boxes: Place one nesting box per 3-4 hens in a dark, quiet corner of the coop for privacy.
6. Install roosting bars higher than nesting boxes to prevent birds from roosting there.
7. Easy Cleaning: To make cleaning easier, design the coop with removable trays or a floor that slopes towards a central drain.
8. Accessibility: Include doors or openings to clean and collect eggs without disturbing the chickens.

Feeding

Proper nutrition is essential for the health and productivity of your poultry. Here is how to feed them properly:

1. Balanced Diet: Feed a balanced diet of commercial chicken feed that meets all of their nutritional requirements.
2. Fresh Water: Maintain a steady supply of fresh, clean water, especially during hot weather.
3. Supplements: Provide grit to aid digestion and oyster shell or eggshell supplements for added calcium.
4. Treats: Serve fruits, vegetables, and grains in moderation. Avoid consuming toxic foods such as chocolate, avocado, and onions.
5. Feeder Design: Choose waste-reducing feeders, like treadle or tube feeders, which also deter pests.

Egg Collection

Efficient egg collection ensures that the eggs are collected cleanly and safely. Here are a few tips:

1. Regular Collection: Collect eggs at least once a day to keep them from becoming dirty or being eaten by the chickens.
2. Clean Nesting Boxes: To ensure clean eggs, keep nesting boxes clean and filled with fresh bedding.
3. Gentle Handling: To avoid cracking and keep eggs as fresh as possible, handle them gently.
4. Storage: Keep eggs at a consistent temperature, either at room temperature for short periods or refrigerated for longer periods.
5. Rotation: Use a first-in, first-out rotation to ensure the oldest eggs are used first.

Following these guidelines will help you create a comfortable and functional environment for your chickens, provide them with the nutrition they require, and efficiently collect and store their eggs. Remember that paying attention to detail and performing regular maintenance is essential for successful chicken keeping. With the proper setup and care, your chickens will thrive, providing you with fresh eggs and the satisfaction of backyard farming.

Backyard chickens can develop a variety of health problems, but with proper care and precautions, many of these can be avoided or reduced. Here are some common health problems and ways to prevent them:

Common Health Issues and How to Prevent them

1. External parasites such as mites, lice, and fleas can infest chickens, leading to irritation and feather loss. Preventive measures include regular coop cleaning, dust baths for chickens, and the use of natural or chemical parasite control. Internal Parasites: Worms can affect chickens' digestive systems. Regular deworming and maintaining clean living

conditions are crucial for prevention.
2. Respiratory Diseases: Chronic respiratory disease and infectious bronchitis can lead to coughing, sneezing, and decreased egg production. Proper ventilation in the coop and avoiding overcrowding can help prevent respiratory problems.
3. Bacterial Diseases: Serious diseases include fowl cholera and salmonellosis. To prevent them, ensure clean water and food sources, implement biosecurity measures to prevent wild birds from accessing the coop, and use available vaccinations.
4. Viral Diseases: Infections such as fowl pox and Marek's disease can quickly spread within a flock. Vaccination is the most effective preventive measure against many viral diseases.
5. Fungal Infections: Moldy bedding or feed can lead to Aspergillosis, a fungal disease. To prevent fungal growth, keep the coop clean and dry, and store feed in airtight containers.
6. Nutritional deficiencies can cause rickets and eggshell problems. To avoid these issues, provide a balanced diet formulated for chickens and supplement with vitamins or minerals as needed.
7. Heat Stress: Chickens can experience heat exhaustion during hot weather. Providing shade, ventilation, and cool water can keep chickens comfortable in hot weather.
8. Egg Laying Problems: Laying hens may experience egg binding or prolapse. A calcium-rich diet and avoiding excessive weight gain in hens can lower the risk of these conditions.
9. Behavioral Issues: - To prevent pecking and cannibalism, provide adequate space, enrich the environment, and establish a pecking order without bullying.
10. Injury: - Accidents or predators can cause injuries in the coop. Secure housing and safe coop design can help reduce the risk of injuries.

Preventive measures include regular health checks for chickens and prompt attention to any issues that arise.

- *Quarantine New Birds: Always quarantine new birds before introducing them into your flock to prevent disease transmission.*
- *Cleanliness: Keep the coop and its surroundings clean and dry to reduce the risk of disease and parasites.*
- *Proper Nutrition: Give your chickens a balanced diet that is appropriate for their age and production level.*
- *Vaccination: To protect yourself from common diseases, follow the recommended vaccination schedule.*

By being vigilant and proactive in your backyard chicken care, you can significantly reduce the likelihood of health problems and ensure a happy, productive flock.

Chapter 5: Generate Your Energy

Chapter 5: Generate Your Energy

Looking At Renewable Energy Options (solar panels, wind turbines)

Exploring alternative energy sources such as solar panels and wind turbines is an excellent approach to help ensure a sustainable future. Both have distinct advantages and can be used for a variety of purposes and locations. Here's a summary of each.

Solar panels use photovoltaic cells to convert sunlight into power. They are silent, unobtrusive, and can be mounted on rooftops or ground. Advantages: Predictable and dependable, especially in places with constant sunlight[2]. - Disadvantages: Energy production can be disrupted by weather, and installation requires significant space. The initial cost of solar panels can be considerable, but they have greatly lowered over time. Standard solar panels have an efficiency of 15% to 25%, but advanced technologies can achieve up to 40%.[2]. Environmental Impact: Solar panels generate clean electricity with zero greenhouse gas emissions during operation. However, manufacturing and recycling have an environmental impact.

Wind turbines create electricity by harnessing the kinetic energy of moving air. They can be erected in a variety of places, including offshore, where stable wind speeds lead to increased efficiency. Wind energy is clean and can generate electricity continuously under optimal conditions. However, wind turbines can be noisy and visually invasive. Wind turbines can pose a risk to birds and bats, and their energy production is intermittent and unpredictable[2]. On average, they operate at 25-30% efficiency, but this can be higher in optimal locations. Similarly to solar, wind power has a low environmental impact during operation, producing no greenhouse gas emissions. The manufacturing and installation processes leave a low carbon footprint[2].

When picking between solar and wind, consider your location, local climate, available space, and energy requirements. Solar panels may be more appropriate for residential areas with limited space, but wind turbines may be a better solution for rural or coastal areas with constant wind patterns. Both technologies are fast improving, becoming more efficient, and less expensive, making them more feasible options for renewable energy generation.

Simple Energy-Efficient DIY Projects

Improving energy efficiency in your house does not always involve a large investment or skilled installation. There are various simple do-it-yourself (DIY) tasks you can do to improve your home's energy efficiency, lower your utility bills, and contribute to environmental conservation. Here's a complete tutorial for some simple DIY projects:

1. Insulate Your Water Heater Tank: Undulated water heaters waste a lot of electricity. Using an insulating blanket can reduce heat loss, lower heating expenses, and boost energy efficiency. This is especially helpful with older models.
2. Insulate Hot Water Pipes: Insulating hot water pipes minimizes heat loss and increases water temperature, allowing for lower water heater settings. Installing foam pipe insulation around pipes is a straightforward project that takes only a few hours to complete.
3. Lower Water Heating Temperature: Setting your water heater to 120°F can conserve electricity and lower the risk of scalding. To make a quick adjustment, simply flip the water heater thermostat.
4. Use caulk to seal air leaks around windows, doors, and other openings. Sealing leaks using caulk is a simple and inexpensive technique to boost your home's energy efficiency.
5. Weatherstrip Windows and Doors: Like caulking, weatherstripping seals the movable components of windows and doors. To prevent air leaks, use materials like foam, rubber, or tape.
6. Install Low-Impassivity Storm Windows: Adding low-impassivity (low-E) storm windows improves insulation and reduces heat loss in winter and gain in summer. They are less expensive than full window replacements and may be installed from both inside and outside.
7. Replace incandescent bulbs with LED lighting for simple and effective energy savings. LED consume at least 75% less energy and last 25 times

longer than standard bulbs.

8. Install programmed Thermostats: A programmed thermostat adjusts the temperature based on your schedule, avoiding heating or cooling an empty house. This can result in big savings on your energy expenses.
9. Insulate Your Attic: Proper attic insulation prevents heat from leaving in winter and entering in summer. This project may involve installing bats or rolls of insulation or adding loose-fill insulation with a rented blower.
10. Create a Windbreak: Planting trees or bushes as a windbreak helps shelter your home from cold winds and reduces heating costs. This is a long-term technique that improves your property's attractiveness.
11. Install Reflective Roofing: In hot climates, reflective roofing helps reduce heat absorption and keep your home cool, minimizing the need for air conditioning.
12. Replace with Energy Star-rated appliances. These are certified to be more energy-efficient and can save you money during the appliance's lifetime.
13. DIY Solar Panels: With some experience, installing solar panels may be a satisfying project. Kits are available with all necessary components, and many countries give incentives to offset costs.
14. Install a Rain Barrel: Using rainwater to irrigate plants or wash your car will save your water cost and conserve a valuable resource.
15. Regular Maintenance: Replace HVAC filters, dust refrigerator coils, and check for leaks to maximize home system performance.

By completing these DIY projects, you can significantly improve your home's energy efficiency. You will not only save money but also contribute to a more sustainable lifestyle. Remember that every little amount counts when it comes to energy saving.

Decreased Dependency On The Grid

Reducing dependency on the grid is one step toward energy independence and sustainability. Here's how you can do it in a few simple steps:

1. Installing solar panels is a highly effective approach to creating electricity. Solar panels, with advances in photovoltaic technology, can dramatically reduce your grid dependence and even allow you to sell extra electricity back to the grid.
2. Wind Turbines: In places with stable wind patterns, small-scale wind turbines can supplement solar panels and supply additional power.
3. Energy Storage: Battery systems can store renewable energy and provide electricity when solar or wind are not available, such as at night or on quiet days.
4. Energy Efficiency: Increasing your home's energy efficiency lowers overall consumption. To reduce energy use, consider updating to LED lighting, insulating your home, utilizing energy-efficient equipment, and being careful of your habits.
5. Smart Thermostats: Programmable thermostats enhance heating and cooling systems, decreasing energy waste.
6. Water Heating: Insulating your water heater and pipework can reduce energy waste.
7. Cooking with Gas: Using natural gas instead of electricity helps lessen dependency on the grid for this regular activity.
8. Natural Lighting and Heating: Use natural light during the day and apply passive solar design concepts to eliminate the demand for artificial lighting and heating.
9. Unplug Devices: Many devices still use power even when turned off. To avoid 'phantom load'[2], unplug them or use smart power strips.
10. Use Manual Tools: Whenever possible, use manual tools instead of electric ones, such as sweeping instead of vacuuming.
11. Alternative Transportation: Using bicycles or walking for short trips can decrease reliance on electric vehicles and reduce grid dependence.

12. Rainwater Harvesting: Collecting and using rainwater for non-potable purposes, such as irrigation, can minimize energy consumption in water supply and treatment.
13. Home Gardening: Growing your food reduces energy use for manufacturing, transit, and storage.
14. Community Energy Projects: Engaging in community solar or wind projects can supply sustainable energy and lessen reliance on the grid.
15. Education and Behavior Change: Educating yourself and your family on energy conservation and implementing deliberate behavioral adjustments can significantly reduce energy consumption.

Implementing these measures will reduce your reliance on the grid, reduce energy bills, and contribute to a more sustainable future. Harnessing renewable energy sources, increasing efficiency, and adopting energy-conscious practices will all contribute to a positive outcome.

Chapter 6: Crafting and DIY Projects.

Chapter 6: Crafting and DIY Projects.

Creating Crafts From Natural Materials (woodworking, weaving, and pottery)

Using natural materials in crafts is an enjoyable and environmentally friendly way to create beautiful and functional products. Whether you prefer carpentry, weaving, or pottery, each craft offers a unique opportunity to connect with nature while also expressing creativity. Here's an in-depth look at these crafts.

Woodworking with natural materials

Woodworking is the technique of making objects from wood. It could range from sculpting miniature miniatures to creating furniture.

1. Wood Selection: Select the proper wood for the job. Hardwoods such as oak and maple are durable for furniture, although softwoods like pine are easier to carve.

- Look for wood with distinct textures or knots to add character.

2. Common tools include saws, chisels, hammers, and sandpaper.

- Advanced tools, such as lathes and routers, can add complexity to projects.

- Techniques vary from simple cutting and sanding to intricate carving and joinery.

3. Finishing: Sand the wood to a smooth finish. - Use natural finishes like linseed oil or beeswax to protect and enhance the wood's natural beauty.

4. Sustainability: - Select reclaimed or responsibly obtained wood.

- Repurpose scraps into smaller projects to save waste.

Woven using natural fibers

Weaving involves the interlacing of threads to create fabrics. It is used to manufacture linen, baskets, and other items.

> 1. Gathering Materials: Natural fibers like cotton, wool, silk, and flax can be spun into yarns for weaving.
> - Common basketry materials include willow, reeds, and bamboo.
> 2. Preparing Fibers: - Clean and card wool to align the fibers.
> - Use a spindle or spinning wheel to turn fibers into yarn.
> 3. Weaving Techniques: - Use a loom to weave yarn into cloth. Looms range from modest frame looms to complex floor looms.
> - Basket weaving does not require a loom and instead involves hand coiling or plaiting materials.
> 4. Natural Dyes: Plant-based dyes like berries, leaves, and bark can be used to color textiles.
> - Experiment with different mordants to stabilize the dye and achieve different colors.

Pottery With Clay

Pottery is one of the oldest crafts, with clay fashioned into pots and other objects.

> 1. Depending on the required finish and firing temperature, choose from earthenware, stoneware, or porcelain clays.
> - If possible, source clay locally to reduce the environmental impact.
> 2. Hand-building techniques include pinching, coiling, and slab construction. - Throwing on a potter's wheel creates symmetrical designs that are ideal for making bowls and cups.
> 3. Decorating and glazing: Use slips (liquid clay) or carve designs into the clay.
> - Use glazes on the pottery to add color and make it waterproof. Natural

glazes can be made from ash and minerals.

4. Firing is the process of solidifying clay in pottery using a kiln. Kilns may be electric, gas, or wood-fired.

- Alternative fire procedures, such as pit firing, use open flames and organic ingredients to provide unique outcomes.

5. Environmental considerations: Recycle clay scraps.

- Use energy-efficient kilns or other firing methods to reduce your carbon footprint.

Finally, using natural materials is a gratifying way to make products that reflect the beauty of the natural world. Whether you're sculpting driftwood, weaving a reed basket, or molding a clay pot, these crafts let you use your hands to create one-of-a-kind products. They also offer the option to embrace sustainability by employing renewable resources and reducing waste. As you study these crafts, you will get a deeper appreciation for the materials and traditional skills that have been passed down through the generations.

Developing Practical Items For Your Household

Making usable products for your farm necessitates a blend of creativity, pragmatism, and sustainability. The goal is to design products that not only serve a purpose but also help your home become more self-sufficient and productive. Here's how you approach it:

1. Identify Your Needs: - Determine which things you frequently use or need for homesteading activities. This could include everything from cooking equipment to gardening tools. - Assess your time-consuming tasks and identify tools or products that can boost productivity.
2. Plan your projects: - Once you've assessed your needs, draw up blueprints or designs for the products you want to manufacture. This could be a simple drawing or a detailed blueprint. Determine the materials you'll need. Whenever possible, select natural, sustainable, or recyclable materials.

3. Gather items: - Search for reusable goods in your home. Recycle old materials like wood, metal, and plastic to make something new. For items such as pottery or textiles, you may need to collect clay or fibers from local merchants or nature.
4. Utilize Traditional Skills: - Learn traditional skills including woodworking, metalworking, weaving, and ceramics. These chores typically require minimal electricity and can be accomplished with hand tools. - To learn these skills, consider taking a class or watching instructional videos.
5. Prioritize quality by carefully constructing each piece. The goal is to create durable, long-lasting products that do not need to be replaced frequently.

- **Quality also includes functionality; ensure that each item performs its intended function properly.**

6. Design Multi-Functional items: - Create items that can provide various functions. Multi-functional items, such as a bench with built-in storage or a pot that can switch from burner to table, can help preserve space and reduce the need for extra items.
7. Test and refine: Use the items you created to find potential improvements. Homesteading requires constant learning and adjustment. - Don't be afraid to modify or rebuild something if it better meets your needs.
8. Share Your Knowledge: If you find a successful design, share it with other homesteaders. This can be blogs, social media, or local workshops[1]. - Sharing information builds a network of self-sufficient people and can provide feedback to help you improve your designs.
9. Prioritize safety when developing and using homestead items. To avoid accidents, utilize protective equipment as needed and follow best practices.
10. Enjoy the process. Remember that creating functional products for your farm should be enjoyable. Take pride in your work and how you're working to create a more sustainable lifestyle.

Building usable things for your homestead not only produces goods but also establishes an independent and resilient lifestyle. Whether it's a sturdy workbench, a set of hand-forged garden tools, or a beautifully crafted quilt, each piece reflects your hard work and the spirit of homesteading.

Decorate Your Home With Handmade Things

Handmade decor is a one-of-a-kind and personalized way to bring warmth, character, and beauty into your home. Handmade objects express a level of craftsmanship and individuality that mass-produced items cannot match. Here's a comprehensive guide to beautifying your home with handcrafted decor.

1. Evaluate Your Space: Begin by evaluating the rooms in your home. Consider the existing color palette, furniture, and overall appearance. Determine which areas could benefit from a bit of homemade flair, such as blank walls, barren shelves, or unsightly corners.
2. Choose a theme or design that expresses your personality and complements your house. This could be a rustic farmhouse, modern minimalism, or vibrant bohemian look. - A unifying theme will guide your crafting decisions and create a pleasant environment.
3. Create Handmade Feature Pieces: - These can be used as focus points in a room. Make a large wall hanging, a statement piece of furniture, or an eye-catching lighting fixture[1]. These things should be visually appealing and set the tone for the overall decor.
4. Use Textural Elements: - Texture adds depth and interest to any environment. Choose handmade fabrics like woven rugs, knitted blankets, or macramé wall artwork. Experiment with yarn, fabric, and natural fibers to create appealing and comfy pieces.
5. Personalize Using Artwork: - Artwork expresses your interests and experiences. Create paintings, sketches, or digital artwork to express your story.

- Showcase your artwork in new ways, such as on a gallery wall or a picture ledge for a more casual look.

6. Craft Functional Decor: Handmade items can be both decorative and functional. Consider making pottery dishes, wooden utensils, or hand-sewn linens[1]. These products provide a purpose and add a personalized touch to everyday tasks.

7. Natural materials such as wood, stone, clay, and plants bring the outside in, providing a calming ambiance ([1]). Use these materials to make planters, coasters, and decorative bowls.

8. Up cycling and Re purposing: - Up cycling is the process of transforming used or abandoned materials into something new and beautiful.

- Up cycle an old ladder into a bookshelf, repurposed jars into vases, or refinish a vintage chair to give it new life.

9. Handmade Lighting: - Lighting sets the tone in a room. To enhance the atmosphere, make your lampshades, candle holders, or string light arrangements.[1]. Create one-of-a-kind lighting designs using materials like paper, glass, or metal.

10. Seasonal Decoration: Use handcrafted decorations to celebrate the changing seasons. To memorialize special occasions, make wreaths, table centerpieces, or Christmas ornaments[1]. - Seasonal decor is easily swapped out to keep your home feeling new and modern.

11. Display Collections: - Place items you collect, such as ceramics, vintage books, or seashells, strategically around your home. Handmade shelves or display cases can be created to showcase your items in their finest light.

12. DIY Wall Treatments: - Instead of wallpaper, use hand-painted murals or stencils. - This method offers complete customization and can be a fun weekend hobby.

13. Handmade embellishments: Even modest homemade embellishments can have a big impact. Decorate your home with homemade items like picture frames, coasters, or ornate knobs to add interest and show off your creativity[1].

14. Engage the Senses:- Decor is more than just visuals. Consider using sensory items like aromatic candles, wind chimes, or a cozy rug. A home that appeals to multiple senses appears more inviting and full.

15. Share and inspire: - Display your created home decor to loved ones. - Your home can inspire others to begin their crafting efforts and discover the

Chapter 6: Crafting and DIY Projects.

joys of handcrafted decor.

By dedicating time to making handcrafted decor, you are not just decorating your home, but also instilling it with memories and purpose. Each piece has a story, whether it's a patchwork fashioned from old T-shirts, a painting inspired by your travels, or a handcrafted table that has hosted innumerable family dinners. Handmade decor is about more than simply aesthetics; it's about creating a home that feels completely your own.

Chapter 7: Herbal Medicine

Growing Therapeutic Herbs

Chapter 7: Herbal Medicine

Growing medicinal herbs may be a wonderfully rewarding pastime that not only beautifies your garden but also provides a steady supply of natural remedies. Whether you're an experienced gardener or just starting, cultivating medicinal herbs offers numerous benefits and is relatively straightforward if you follow a few basic guidelines.

Choosing Your Herbs

The first step in cultivating medicinal herbs is to choose which ones to grow. Consider the weather, the available space, and your individual health needs or interests. Some typical therapeutic herbs that are easy to grow are lavender, chamomile, mint, sage, and thyme. Each herb has various needs and benefits, so it's important to research them before planting.

Understanding the Environment

Herbs prefer sunny conditions and well-drained soil. Certain plants, however, may thrive in partial shadow or require more rain. It is vital to research each herb's native environment to replicate those conditions as closely as possible in your garden.

Soil Preparation

Preparing the soil is essential for the health of your herbs. Compost can help enhance the structure and nutrient content of your soil, and most plants prefer a pH range of neutral to slightly alkaline. If you're planting in pots, choose a high-quality potting mix with adequate drainage.

Planting

You can start herbs from seeds or purchase young plants from a nursery. If you're starting from seed, adhere to the depth and spacing guidelines on the packet. Gently remove young plants from their containers, pick off any pot-bound roots, and plant them at the same depth as in the pot.

Watering

Each herb requires different amounts of water. In general, deep yet infrequent watering promotes good root growth. Overwatering can lead to root rot, so ensure the soil is dry to the touch before watering again.

Feeding: Most plants don't need much fertilizer. Overfeeding may produce lush foliage with diminished flavor and therapeutic efficacy. If necessary, use a balanced organic fertilizer sparingly.

Pruning and Harvesting

Regular pruning encourage bushy growth and prevents herbs from becoming lanky. Harvest herbs in the morning, after the dew has dried but before the sun reaches its highest, when the essential oils are most concentrated. Always leave enough leaves so that the plant can continue to grow.

Pest and Disease Control

Healthy, well-cared-for plants are less susceptible to pests and diseases. If problems arise, use organic methods of control whenever possible, such as introducing beneficial insects, using neem oil, or removing diseased plant parts.

Winter Care

Some herbs are perennial, meaning they return year after year, while others are annual or biennial. To protect perennial herbs from cold temperatures in the winter, they may need to be mulched. Annuals and biennials must be changed annually.

Using Your Herbs

Once harvested, your herbs can be either fresh or dried for later use. Drying is a simple process that involves hanging bunches of herbs in a warm, airy room away from direct sunlight. Once dry, store them in airtight containers away from light and heat.

Conclusion

Growing medicinal plants benefits both your health and the environment. It increases biodiversity, benefits pollinators, and can be done naturally without the use of hazardous pesticides. With time and care, your medicinal herb garden will thrive, providing beauty and natural medicines for many years to come.

Herbal Remedies For Common Illnesses

Herbal remedies have been used for centuries to treat a variety of common ailments. These natural therapies are derived from plants and have been known for their therapeutic capabilities; they are usually used as a safer and less expensive alternative to medications. Here, we'll look at some of the most well-known herbal remedies and how they might be used to treat common health issues.

Echinacea
Echinacea is widely used to both prevent and treat common colds. It is considered to boost the immune system and reduce the symptoms of colds and other illnesses. According to studies, echinacea can help shorten and milden a cold.

Ginseng
Ginseng is another immune booster recognized for its rejuvenating properties. It is commonly used to relieve fatigue and is believed to improve mental clarity and physical endurance. Ginseng has also been used to treat diabetes and sexual dysfunction in men.

Peppermint
Peppermint is well-known for its cooling properties and refreshing aroma. It is frequently used to treat digestive issues such as bloating, gas, and indigestion. Apply peppermint oil to your temples to ease headaches.

Chamomile
Chamomile is a gentle herb that is widely used to promote relaxation and sleep. It has anti-inflammatory and antispasmodic properties, which make it useful for treating stomach cramps and other gastrointestinal discomforts. Chamomile tea can also aid in soothing sore throats and nerves.

Lavender
Lavender is known for its calming and sedative effects. It is used to manage stress, anxiety, and insomnia. Lavender oil's antibacterial and anti-inflammatory properties make it effective for treating skin wounds, burns, and bug bites.

Turmeric
Turmeric contains curcumin, which has strong anti-inflammatory and antioxidant properties. It is used to treat a variety of conditions, such as arthritis, heartburn, and joint pain. Turmeric is also being studied for its potential to prevent and treat Alzheimer's and cancer.

Ginger
Ginger relieves nausea and vomiting, making it a popular treatment for

motion sickness and morning sickness during pregnancy. It also has anti-inflammatory properties and may aid with osteoarthritis pain.

Licorice Root

Licorice root has a long history of use in treating gastrointestinal problems. It can help with stomach ulcers, heartburn, and colic. Licorice is also used to treat a sore throat and cough.

St. John's Wort

St. John's Wort is commonly used to alleviate mild to moderate depression. It is considered to work by increasing serotonin levels in the brain, hence improving mood and reducing depression symptoms.

Valerian Root

Valerian root is another herb noted for its sedative properties. It is widely used to treat sleep disorders, notably insomnia. Valerian is also used to relieve anxiety and stress.

Tea Tree Oil

Tea tree oil contains antibacterial properties and is used to treat a variety of skin conditions, including acne, athlete's foot, and dandruff. It also acts as a natural disinfectant against cuts and scrapes.

Aloe Vera

Aloe vera is well known for its skin-healing properties. It is applied topically to treat sunburn, minor burns, and skin irritation. Because of its moisturizing characteristics, aloe vera gel is often used in cosmetics.

Milk Thistle

Milk thistle is used to protect and optimize liver function. It is supposed to cleanse the body and is used to treat liver diseases like hepatitis and cirrhosis.

Fever few

Fever few is used to prevent migraines and reduce their frequency. It has anti-inflammatory properties and is used to treat fever and arthritis.

Saw Palmetto

Saw palmetto is widely used to treat benign prostatic hyperplasia (BPH), which causes an enlarged prostate gland in men. It is supposed to aid with the urinary symptoms associated with BPH.

Conclusion

Herbal medicines offer a natural approach to health and wellness, treating a wide range of common ailments with fewer side effects than traditional medications. While many herbs are safe and helpful, it is crucial to consult with a healthcare practitioner before beginning any herbal treatment, especially if you are pregnant, nursing, or using other medications. With the proper knowledge and application, herbal remedies can be a valuable addition to your medical regimen, providing relief and assisting recovery gently and naturally.

Harvesting and preparing herbal teas, tinctures, and salves

As a dedicated herbalist, I like gathering and preparing my herbal medicines. The approach is both an art and a science, requiring patience, precision, and a deep respect for the plants. Allow me to demonstrate how I collect and prepare herbal teas, tinctures, and salves while teaching knowledge of the earth's natural healing capabilities.

Harvesting Herbs

My voyage begins in the garden, where I methodically select the best time to harvest each herb. I prefer a dry morning after the dew has evaporated but before the midday sun depletes the plants' essential oils. I cut the herbs with clean, sharp scissors, taking care not to damage the plant or remove more than one-third of its growth. This ensures that the plant is healthy and can continue to produce.

Drying Herbs for Tea

After harvesting, I knot the herbs into small bundles and hang them upside down in a warm, open room away from direct sunlight. This slow-drying process preserves the herbs' medicinal characteristics. When the leaves and blossoms have completely dried, I smash them, remove the stems, and store them in airtight containers. To make tea, I steep a teaspoon of dried herb in a cup of boiling water for around 10 minutes before draining and drinking.

Creating Tinctures

Tinctures are very concentrated botanical extracts. I begin by filling a jar with freshly cut herbs, then add enough vodka or brandy to completely cover the plant material. Over a few weeks, the alcohol extracts the active compounds from the herbs. I shake the jar every day to help in the process. After four or six weeks, I strain the liquid through a fine mesh sieve coated with cheesecloth, removing as much liquid as possible. The completed tincture is stored in dark glass dropper bottles labeled with the plant name and date.

Making Salve

Slaves are good for topical use. I start by infusing oil with herbs. I slowly heat a carrier oil, such as olive or coconut oil, and then add dried herbs, simmering them for a few hours on low heat. After purifying the oil, I measure it, add beeswax, and heat it until it melts. For each cup of infused oil, I use around 1/4 cup of beeswax. After blending, I spoon the mixture into tins or jars and let it cool. As it sets, I'll have a soothing salve ready for cuts, bruises, and dry skin.

Ritual of Preparation

For me, the whole harvesting and preparation process is a ritual. It is a means of connecting with nature and harnessing its therapeutic abilities. I take pleasure in knowing that the remedies I create are pure, effective, and carefully made. Whether it's a calming cup of chamomile tea, an energizing ginseng tincture, or a healing calendula salve, each embodies the essence of the plant and the care I put into its creation.

Sharing the harvest.

I routinely share my herbal concoctions with friends and family, spreading the word about the benefits of these natural remedies. It's a way to build community and promote a comprehensive approach to health. Each salve jar, tincture bottle, and tea pouch embodies the power of plants and the herbal medicine tradition.

Reflection and Respect

In my years of experience, I've realized that the key to successful herbal preparation is not only technique but also respect for the plants and the environment. It's about understanding natural growth cycles and enjoying what nature has to offer. As I create each remedy, I reflect on this relationship and my responsibility to use these herbs wisely and ethically.

Conclusion

Harvesting and preparing herbal teas, tinctures, and salves is a gratifying experience that reconnects me with nature and promotes my well-being. It's a technique that I've honed over time, and each batch provides a new opportunity to learn and grow. I advise everyone interested in herbalism to start their journey and experience the satisfaction of making something healthy and wholesome with their own hands. It's a path that leads to better health, greater understanding, and a more sustainable way of living.

Chapter 8: Raising Small Livestock

Introduce Honeybees, Bunnies, And Goat

Chapter 8: Raising Small Livestock

Raising animals like bees, bunnies, or goats may be a fun and rewarding activity. Each of these species has unique advantages that necessitate various levels of care and money. Let us see what it is like to keep each of them.

Bees: The Sweet Buzz of Beekeeping

Keeping bees is a fantastic activity that not only produces fresh honey but also benefits your local ecosystem by pollinating. Here's all you need to know about beekeeping:

- Hives: Your bees will reside in hives, which are structures that store honey and raise their offspring.

- Supplies: You'll need protective gear, a smoker to keep the bees calm, and tools to work with the frames inside the hive.

- Considerations: To effectively maintain the hive, you must monitor for illnesses and pests and understand their life cycles.

Rabbits Are Furry Friends of the Homestead.

Rabbits are a popular choice for small-scale livestock since they require little maintenance and can produce meat, fur, and even garden waste. Here's the deal with rabbit raising:

- Housing: Rabbits demand a clean, dry, well-ventilated environment. Hutches and cages are popular alternatives.

- Diet: They eat a wide range of meals, including hay, fresh vegetables, and specially formulated pellets.

- Breeding: Rabbits breed easily and have short gestation periods, allowing you to quickly grow your rabbitry as needed.

Goats: Versatile and Entertaining Livestock.

Goats are extremely versatile animals that may provide milk, meat, fiber, and even companionship. They are known for their active and inquisitive nature. Here's what goat-keeping entails.

- Space: Goats need room to wander and graze, as well as shelter from the elements.

- Diet: They are browsers who prefer to eat leaves, twigs, and shrubs, but they also need a well-balanced diet high in hay and grains.

- *Milk Production: If you want to raise goats for milk, you'll need to breed them first and then milk them regularly.*
- *Keep at least two goats for companionship, as they flourish in groups.*

In all cases, it is vital to undertake extensive research and be well prepared before introducing cattle into your life. Check your local regulations because each animal species has distinct needs and legal requirements. Bees, rabbits, and goats can be great additions to your farm, whether you're looking for honey, fur, or milk, or simply enjoy caring for these amazing critters.

Benefits of tiny livestock

Raising tiny cattle is an Eco-friendly and effective way to improve your homestead or small farm. These small animals often require less space, use fewer resources, and are easier to manage than larger cattle. Here are some of the main benefits of having little livestock:

1. Efficient Land Use:

Small livestock are ideal for individuals with little land. They thrive in small pastures or backyards, making them ideal for urban or suburban farmers.

2. Lower costs:

Small livestock are generally less expensive to purchase and maintain. They consume less feed and water than larger animals, leading to lower ongoing care costs.

3. Diversified income

Small cattle can provide eggs, milk, meat, wool, and honey, creating many income streams and self-sufficiency.

4. Easy management:

Small cattle are easier to manage and suitable for novices or those with physical limitations.

5. Soil Health:

Small livestock dung increases soil fertility and structure, resulting in enhanced crop yield on homesteads.

6. Sustainable Food Production:

Small cattle may produce nutritious food from nonhuman sources such as grass and kitchen waste.

7. Community and Heritage Preservation:

Raising heritage breeds of small animals promotes genetic diversity and traditional farming practices.

8. Educational Opportunities:

Keeping little cattle provides families and communities with educational opportunities that emphasize responsibility and food origins.

9. Environmental Benefits:

Small livestock has a smaller environmental impact than larger animals, promoting sustainable farming practices.

10. Personal Satisfaction:

Caring for animals can provide joy and satisfaction, leading to a higher quality of life.

To summarize, small livestock give a practical and rewarding way to participate in animal husbandry, even on a limited scale. They can help you live a more sustainable lifestyle, provide a diverse range of products, and add enjoyment to your daily routine.

Basic care and management

Basic livestock care and management are critical to maintaining a farm healthy and productive. Whether you're raising chickens, goats, cattle, or any other farm animal, there are some universally relevant tactics for keeping them healthy. This thorough book will help you grasp the essentials of cattle care and management.

Shelter & Housing

Livestock requires refuge from inclement weather and predators, as well as a comfortable resting area.

Shelter design should consider species-specific requirements, such as room, ventilation, and insulation.

- Shelters need frequent cleaning and maintenance to prevent manure and muck buildup, which can lead to health issues.

Nutrition & Food

- Balanced Diet: A nutritious, well-balanced diet is essential for your animals' health and productivity. This includes grass, cereals, and supplements tailored to their nutritional requirements.

- Clean Water: Access to safe, pure water is essential. Water is crucial for digestion and overall health, particularly in lactating animals.

- Maintain a consistent food plan for better digestive health and nutrition throughout the day.

Healthcare and Veterinary Assistance

- Schedule regular veterinary check-ups to monitor livestock health and detect abnormalities early.

- Vaccinations: Follow a vaccination schedule to prevent your animals from common diseases and infections.

Implement a parasite control program to address internal and external parasites that might negatively impact animal health.

Handling and Behavior

- Gentle Handling: Proper handling skills lessen animal stress and danger of harm for both livestock and handler.

- Identify typical and abnormal behavior to evaluate health, comfort, and well-being.

- Improve animal safety and handling by training them in routine management methods.

Reproduction and Breeding

- Develop a breeding strategy based on genetics, health, and your livestock operation's objectives.

Pregnant and newborn animals need special care in a clean, safe environment.

- Keep detailed records of breeding, births, and health treatments for effective herd management.

Pasture and Land Management.

Rotational grazing promotes pasture health and offers nutritious food for animals.

- Secure fencing is necessary to keep animals secure and contained in their grazing areas.
- Proper manure management boosts soil fertility and reduces environmental contamination.

Bio security

To prevent disease spread on your farm, employ biosecurity measures.

- Implement a quarantine protocol for new or sick animals to safeguard the rest of the herd[1].
- Regularly clean and disinfect equipment and facilities to ensure a sanitary environment.

Emergency Preparedness

- Create a strategy to protect your animals during natural disasters, including resources and processes.
- First Aid: Keep a well-stocked first aid bag and know basic animal first aid methods.
- Identification of all animals is essential for recovery operations in case of theft or escape.

Education & Resources

- Lifelong Learning: Stay current on cattle care and management techniques through workshops, courses, and reading materials.
- Networking: Meet farmers and industry professionals to exchange knowledge and experiences.

To recap, basic livestock care and management entails a wide range of actions that are vital to your animals' health and output. You can ensure that your cattle thrive by giving appropriate shelter, food, and medical treatment, as well as competent environmental management. Remember that the key to successful livestock management is attention to detail, consistency, and concern for your animal's well-being.

The Best Rotational Grazing

Rotational grazing is a planned method of livestock management that involves moving animals between pastures or paddocks. This enables you to enhance forage utilization, improve pasture health, and promote long-term land management. Let's have a look at some effective rotational grazing techniques.

1. Divide the pastures into paddocks.
- Create smaller grazing zones by dividing your pasture into paddocks. Paddock sizes and shapes can vary depending on grain supply, herd size, and site layout[2].
- Smaller paddocks provide more accurate control over grazing intensity and vegetation regrowth.

2. Rotate the livestock regularly.
- Transfer your animals between paddocks at regular intervals. A variety of factors influence rotation frequency, including grass growth rate and herd number.
- To prevent overgrazing, give each paddock adequate time to recover before grazing again.

3. Monitor forage growth:
- Monitor fodder growth in each paddock. When the grass reaches grazing height, rotate the livestock.
- Understanding grass growth is essential for proper rotational grazing. Grasses have three stages of development: vegetative, transitional, and reproductive. For optimal feed quality and quantity, graze during the transitional period when grasses are most nutritious.

4. Provide adequate water sources.

- Ensure that each paddock has access to clean water. Proper hydration is essential for cow health and productivity.

- Livestock water requirements can be met with portable or fixed water tubs linked to main pipelines.

5. Adjust the stocking rates:

- Calculate an appropriate stocking rate based on your land's carrying capacity and available forage.

Overstocking can lead to excessive grazing and soil damage, whereas under stocking might result in underutilized forage.

6. Maintain the fencing.

- Proper fencing is required for rotational grazing. Good perimeter fencing prevents livestock from straying into adjacent paddocks.

Consider using virtual or electric fencing to simply divide pastures.

7. Observe the soil health:

- Perform frequent soil health assessments in each paddock. Healthy soil encourages better forage growth and nutrient cycling.

Rotational grazing enhances soil structure and organic matter while reducing compaction.

8. Prepare for seasonal changes.

- Adjust your rotational grazing approach in response to seasonal variations in fodder growth and weather conditions.

To prevent overgrazing during rapid grass growth seasons (e.g., spring), limit grazing duration in each paddock.

9. Consider biodiversity.

- Rotational grazing can enhance biodiversity by creating diversified habitats for a wide range of plant species, insects, and birds.

Combining grasses, legumes, and forbs improves pasture health and resilience.

10. Record Keeping:

- Maintain records for your rotational grazing system. Note which paddocks were grazed, when, and for how long.

- This information helps track forage utilization, plan future rotations, and evaluate management effectiveness.

Remember that successful rotational grazing requires flexibility and adaptability. Each farm is unique, and factors such as climate, soil type, and livestock species will influence your strategy. Implementing these best practices can help you maximize forage utilization, improve soil health, and create a long-term grazing system for your animals.

Conclusion

- Celebrating tiny victories along your homesteading journey.

Celebrating small victories is a crucial part of remaining motivated and happy on your homesteading journey. Homesteading can be difficult, but recognizing and celebrating minor accomplishments along the way can boost morale and encourage future progress. Here's a complete description of how to commemorate these occasions.

Recognize every step forward.

Every step toward homesteading, however small, shows progress. The first sprout in your garden, the successful canning of your produce, or the first egg laid by your fowl are all significant events. Recognize your accomplishments by recording them in a notebook or sharing them with friends and family.

Share your success

Social media networks and homesteading forums are excellent places to share your accomplishments. This not only allows you to chronicle your accomplishments, but it also connects you to a community that recognizes and appreciates your efforts.

Reward yourself.

Create a rewards system for yourself. For example, after successfully harvesting your crops, reward yourself with a delicious supper created from your produce. Alternatively, after a day of hard work, reward yourself with an evening of relaxation.

Educational Investment

Invest in your homesteading education to celebrate. Attend a class, buy a book on a new homesteading skill, or sign up for an online course. This not only pays you but also enhances your homesteading abilities.

Prepare a farm-to-table supper.

Invite friends and neighbors over for supper made from your homestead's bounty. This is a fantastic way to celebrate your achievements and share the joy with others.

Create a Victory Garden.

Set aside a little section of your garden to celebrate your accomplishments. Plant flowers or specific herbs here to create a space of beauty and reflection on your successes.

Document your progress.

Capture photographs of your homesteading activities and create a visual timeline to track your progress. This might be a useful method for reflecting on your development and celebrating victories.

Create Something from Your Homestead

Make something unique with stuff from your own home. Whether it's making a wreath out of garden herbs or building furniture out of wood on your property, doing something with your hands is a rewarding way to celebrate.

Take some time to reflect.

Schedule frequent time to reflect on your accomplishments. This can be a tranquil moment with a cup of tea at dawn, or a walk around your property to acknowledge all of your hard work.

Set new goals.

With each victory, set a new goal. This sustains momentum and gives you fresh accomplishments to look forward to. It is both a celebration of the past and a look forward to the future.

Involve Your Family

Make sure to involve your family in the festivities. Homesteading is often a family project, and celebrating accomplishments together strengthens bonds and spreads joy.

Give back to the community.

Consider giving back as a way to celebrate. Donate extra vegetables to a local food bank, or offer to help a neighbor with their homesteading efforts. This encourages the spirit of homesteading and recognizes your abundance.

Take the day off.

Sometimes the best way to celebrate is to take a well-deserved break. Set aside a day from homesteading tasks to do something you enjoy, such as hiking, reading, or simply relaxing.

Personalize your celebrations.

Remember that celebrations are personal. What you consider a victory may be different for someone else. Celebrate in ways that bring you joy and satisfaction.

To summarize, celebrating minor victories on your homesteading journey entails acknowledging progress, sharing your accomplishments, rewarding yourself, and setting new objectives. It serves as a way to maintain your enthusiasm and inspiration while also reminding you of the hard work and dedication required for homesteading. By taking the time to celebrate, you honor your accomplishments and prepare to face the next challenge with renewed energy.

Encourage newcomers to learn, adapt, and enjoy the experience

Homesteading can be a rewarding experience, with its own set of challenges and successes. If you're a newcomer, here's a detailed motivation to learn, adapt, and enjoy the process.

Embrace the learning curve.

Homesteading is a continuous learning process. Every day brings new learning opportunities, whether it's about raising cattle, growing vegetables, or conserving food. Embrace the learning curve with enthusiasm. Remember, every master was once a beginner.

Begin small.

Don't try to complete everything at once. Start with a small garden or a few chickens. Small successes will gradually build your confidence and skills.

Chapter 8: Raising Small Livestock

Celebrate your successes!

Make time to recognize your achievements, no matter how minor. Did your first tomato crop flourish? That's a victory! These celebrations will rekindle your passion for homesteading.

Connect with the community.

Join a homesteading club, whether it's local or online. The community provides invaluable information and assistance. They can offer guidance, resources, and encouragement when you need it the most.

Be Patient with Yourself.

Mistakes are inevitable, but they are also invaluable. Be patient and gentle with yourself when things don't go as planned. Each error serves as a stepping stone to a deeper understanding.

Adapt and innovate.

Homesteading requires adaptability. If a crop fails, take advantage of the opportunity to research and experiment with new methods. A successful homestead relies heavily on innovation.

Keep a journal.

Keep track of your journey. A journal can be a useful record of what works and what doesn't, as well as a source of inspiration as you consider how far you've come.

Nature can teach us.

Nature provides an excellent lesson for a homesteader. Study seasonal patterns, animal behavior, and plant growth. Nature contains knowledge that can help you with your homesteading efforts.

Invest in high-quality tools.

Quality tools can make a significant difference. Invest in the best tools you can afford; they will make your work more efficient and enjoyable.

Prioritize sustainability.

Homesteading entails living sustainably. Learn about permaculture, composting, and water conservation. These methods not only benefit the environment but also boost your homestead's self-sufficiency.

Take breaks.

Don't forget to rest. Homesteading is a physically demanding endeavor that can lead to burnout. Taking a break allows you to approach your tasks with new energy and perspective.

Stay curious.

Always remain curious and open to new ideas. Read books, attend workshops, and watch tutorials. The more you understand, the more you can do with your property.

Enjoy the process.

Finally, remember to enjoy the experience. Homesteading is about the journey as much as the outcome. Take pleasure in your everyday routines, quiet moments with nature, and the satisfaction of self-sufficiency.

Homesteading is more than just self-sufficiency; it is a way of life that promotes learning, growth, and a strong connection to the land. As a starter, you can shape your homestead to represent your ideals and aspirations. So go forth, learn excitedly, adapt creatively, and savor every moment of this wonderful journey.

www.ingramcontent.com/pod-product-compliance
Lightning Source LLC
Chambersburg PA
CBHW050235230526
45470CB00005B/1961